Gift From The Sea

How a Protein from Jellyfish Fights the Aging Process

by

Mark Y. Underwood

AuthorHouse™
1663 Liberty Drive, Suite 200
Bloomington, IN 47403
www.authorhouse.com
Phone: 1-800-839-8640

The ideas, procedures, and suggestions herein are not intended to replace the services of a trained health professional. If you have any pre-existing medical conditions, consult your physician before adopting any of the suggestions and/ or procedures in this book. Any use of the information provided in this book is at the reader's discretion. If you are taking prescription medication on a regular basis, please check with your physician before using Prevagen™. The statements contained within this book have not been evaluated by the Food and Drug Administration. The products mentioned are not intended to diagnose, treat, cure, or prevent any disease.

© 2007 Mark Y. Underwood. All rights reserved.

No part of this book may be reproduced, stored in a retrieval system, or transmitted by any means without the written permission of the author.

First published by AuthorHouse   8/28/2007

ISBN: 978-1-4343-3125-0 (sc)

Printed in the United States of America
Bloomington, Indiana

This book is printed on acid-free paper.

Contents

Dedication	vii
Acknowledgements	ix
Introduction	i
Chapter 1. The Quincy Bioscience Story	**1**
Building a Company	5
Strong Science	7
Chapter 2. Getting Old Doesn't Have to Be a Bummer	**9**
Soaring Costs	11
The Role of Calcium in Aging	12
Calcium's Link to Disease	14
Protecting Brain Cells with Prevagen™	15
Chapter 3. The Science Behind Prevagen	**19**
The Testing Begins	20
History of Aequorin	26
The Safety of Prevagen™	31
Chapter 4. The Horizon	**35**
Boomers and Beyond	35
The Diseases of Aging	38
Remember the Calcium Connection!	42
Taking on the "Silent Killer"	43
Fighting Inflammation	45
The Promise of Aequorin	46
Selected References	49

Dedication

This book is dedicated to the many who pray for hope and health and peace. May all their prayers be answered. In addition to my Savior above, of whom I hope you all would know, I would like to thank my dear wife Sarah, who has endured the years at my side, unfailing in her confidence and faith. We are truly one and the world would be a lonely place for me without her in my life. Thank you, dear.

Acknowledgements

In life's journey there exist many unanswerable questions. I am encouraged to note the following excerpt from Scripture and meditate upon its meaning in the context of my life's pursuit.

John 9:1-3 Jesus Heals a Man Born Blind

> 1 As he went along, he saw a man blind from birth. 2 His disciples asked him, "Rabbi, who sinned, this man or his parents, that he was born blind?"
>
> 3 "Neither this man nor his parents sinned," said Jesus, "but this happened so that the work of God might be displayed in his life. 4 As long as it is day, we must do the work of him who sent me. Night is coming, when no one can work. 5 While I am in the world, I am the light of the world."

Sometimes the difficult things in life exist for the sole purpose of pointing us in another direction. Sometimes the difficult things in life, like illness, contain within them hidden values that are often unseen or unappreciated. I truly believe that this Gift from the Sea is even more a Gift from Above.

I also have a favorite quote I would like to share:

> "There is no limit to what a man can do, or where he can go, if he doesn't mind who gets the credit."
>
> —*Ronald Reagan, First Inaugural Address,*
>
> *January 20, 1981*

This book contains only my account of some of the events related to the development of our technology, and only a small portion of the truly fantastic details I have experienced on this journey. I am sure there will be many more experiences to document from this day forward. We have a very bright future and I am certain the many details that are not captured in this work will be amazing beyond belief.

There exist countless teachers and encouragers I have encountered along the way, and to them I am greatly indebted. Our team at Quincy Bioscience is truly remarkable and, without them, this program would indeed only exist on paper, and would not have sprung into life as it has. The faith of Michael Beaman has been a constant encouragement and we have shared many trials and victories together. I could not be blessed with a better business partner, or team, with which to share this journey.

Many have asked to hear this story, and because it has offered them encouragement in their lives, we have recorded it in this form with the hope that this book offers a witness of these events as an encouragement to others. There are indeed reasons to have hope for health and peace, and a reason to have Faith.

Introduction

Meet Mark Underwood

Mark Underwood is the president and one of the founders of Quincy Bioscience and the remarkable man behind the story of jellyfish and the treatment of neurodegenerative diseases and aging.

It's been called one of the best places in the United States to live and work. This is Madison, Wisconsin, the capital of the state and a biotech hotspot—home to University of Wisconsin and the headquarters of Quincy Bioscience, a promising young biotechnology company that is set on eradicating debilitating neurological conditions such as Alzheimer's and Parkinson's diseases.

A clue to what is happening inside Quincy's doors may be found in the president's office. On his desk is a simple leather plaque. "It Can Be Done" is the message embossed on the front in gold lettering. This is the prevailing attitude and the motto at Quincy Bioscience, and can even be found on their business cards. Hanging on the wall directly across from the desk is a print of a picture of the Swedish sculpture entitled "Hand of God" by Carl Milles.

It is the unique combination of these two images—the sheer determination of the four-word statement and the concept that man's place and purpose should be recognized as being within the hand of God—that really states the true personality of the company and one of the company's founders, Mark Underwood. Without these two complementary perspectives, this story would never really need to be told.

With perspective, inspiration and innovation sometimes happen in plain sight where others hadn't gone before. And so, if it hadn't been for the unique way Quincy Bioscience envisioned jellyfish research, Prevagen™ might not exist today.

If you haven't heard of Prevagen, Quincy's groundbreaking supplement that recently became available in the marketplace, you will soon. It's anticipated Prevagen will soon be well known throughout the world.

This global recognition will be due in part to the fact that Prevagen is a product of "firsts." It is the first calcium-binding protein supplement ever made, the first anti-aging supplement with strong scientific evidence to back it up, and the first dietary supplement derived from jellyfish.

Mark Underwood is the man who discovered the key to reducing cellular death by using a natural source—a jellyfish protein called aequorin. His breakthrough discovery and application is, as Mark likes to point out, "a gift from the sea."

He learned about the calcium-binding properties of aequorin, a protein excreted from a specific species of jellyfish (*Aequorea victoria*), over ten years ago, but the scientific research behind it goes back to the 1960s. Researchers before Mark were interested in jellyfish mainly because they were curious why they glowed in the dark.

During the next thirty years aequorin was safely used as a laboratory marker because of its unique luminescent properties.

When Mark Underwood entered the picture, he had something else in mind. Three decades after scientists first used aequorin in laboratories, Mark saw what others had overlooked. He had a unique idea that no one before him had visualized. Mark was able to grasp the true value of jellyfish protein in the fight against aging.

If you think calcium is only found in the bones, you're almost right. About 99 percent of calcium in the body is in the bones, but about one percent is found in the nervous system, which controls brain function.

The brain and nerves are also the targets of neurological diseases.

When Mark was a college student at the University of Wisconsin-Milwaukee, he became intrigued with finding out how aequorin might protect brain cells, and through that process, guard against neurodegenerative conditions like Alzheimer's, Parkinson's, and stroke.

Mark began recording copious notes about his vision of using jellyfish protein to slow down memory loss and aging, an idea that seemed too good to be true. Along the way, while he pursued other endeavors in the business world, he could not shake off his idea that jellyfish could play a key role in slowing the aging process by using aequorin as a supplement—after all, it had long been documented in the medical community that the depletion of calcium-binding proteins was associated with aging and the onset of neurological diseases. Why not just add these valuable proteins back into the body?

Mark has experienced many fortuitous moments in his career, from the time he graduated from UW-Milwaukee in 1996, to the formation of Quincy Bioscience, and most recently the market launch of the aequorin-based supplement, Prevagen, on September 1, 2007. He has continually "steered the course," even when skeptics challenged his passion and zeal.

Mark Y. Underwood

Mark has unshaken confidence in Prevagen because of the strong science that stands behind it.

In addition to solid scientific research, Mark has experienced several intriguing coincidences or unexplainable "signs" over the past decade. For Mark, these were confirmations that showed Quincy Bioscience was on the right path in developing and commercializing a calcium-binding protein from jellyfish.

When asked how he had the foresight and knowledge to envision using aequorin, the special protein that comes from jellyfish, to fight aging, Mark Underwood answers with characteristic humor and humility, "I didn't design it; I just looked under the right rock."

Read on to find out more about the fascinating world of Mark Underwood, whose wisdom and vision will help people around the world age more gracefully, and with greater health and dignity.

Chapter 1.
The Quincy Bioscience Story

As a college student at the University of Wisconsin-Milwaukee, there was a time when I was serious about becoming a neurosurgeon. My interest in brain science was motivated by the fact that my mother has multiple sclerosis —I was keenly interested in learning how that disease process worked. But I also loved taking ideas and designing things around them. That is what changed my focus to neurochemistry. I felt it was possible to help more people successfully by adjusting brain chemistry through natural means, with supplementation or naturally-derived medicines, rather than surgery.

I took that interest to the next level after reading a case study about a patient who contracted the symptoms of Guillain Barré syndrome after a jellyfish sting, and was successfully treated with calcium channel blockers. Guillain Barré syndrome creates multiple sclerosis-like symptoms that affect the peripheral nervous system. What really fascinated me was that this person's Guillain Barré symptoms improved to some extent from the treatment, but also that the jellyfish could inflict this problem, yet not suffer from it within its own simple nervous system. To me, that raised many questions: What is calcium's role in damaging the nervous system? Can mediating calcium fight neurological conditions? How can jellyfish deliver a sting and not suffer any neuropathy? Why doesn't a jellyfish poison itself?

I began to research these remarkable creatures and the role of calcium in the nervous system. They are very simple organisms. If you take away all the higher functions of thought from the human brain, you really end up with a very simple nervous system—as simple as that of the jellyfish. I wanted to know how their design and function could help fight crippling diseases like multiple sclerosis. I began to take notes about a unique jellyfish protein called aequorin and how it might help fight these diseases. I felt it had even bigger potential than previous researchers had recognized as a simple fluorescing marker used in the lab. The connection between health and calcium regulation in the body had not been missed—but could a jellyfish protein play an integral role in health? It was an idea that had not been considered, even though calcium is the underlying electricity that runs through our bodies.

> If you take away all the higher functions of thought from the human brain, you end up with a very simple nervous system—as simple as that of the jellyfish.

Between 1994 and 1996, when I was in college, my curiosity about aequorin led me to write "The Aequorin Hypothesis," a 60-page research paper that attracted the interest of graduate advisors in the clinical laboratory sciences department. Although I was offered a graduate position, a sponsoring professor who liked my ideas left the department to become an academic dean at Clemson University, just as my research was about to begin in the lab. It was unfortunate timing, and instead of academia, I resolved to start my career in industry.

Over the years outside the lab, as I learned more about aequorin, I kept coming back to this idea: Instead of blocking calcium channels, what if we let calcium flow into brain cells and then manage it, in the most natural way possible? Why had no one else thought of this approach?

Sometimes innocence, coupled with curiosity and a new perspective, plants the seeds for a new discovery. That is exactly what happened here—perhaps I was naive enough to think about things differently, but I continued to ponder the question, "What is the role of calcium in the nervous system?" This was before the advent of Internet search engines, so I could not just go online and type in "aequorin." There were many long evenings spent in the library of the closest medical college.

I have long believed that things happen for a reason; I do not believe in fate, but purpose. Over the course of the last ten years I have experienced many things that seemed to underscore the fact that we were on the right path in developing Prevagen. These events gave me hope along the way, and underlined the science behind the ideas I had were just as compelling. As I dug deeper and learned more, the project made more and more sense.

One of these events occurred during a volleyball game after church, when one of the players introduced himself as a molecular biologist. I told him I was an amateur biologist, joking that I had an "odd" interest in jellyfish proteins. When he further explained he had just finished a seminar on the east coast learning to work with aequorin, I was rather shocked! What were the odds I would meet, during a volleyball game, one of the few people in the world who knows how to make this remarkable protein? How many people knew that jellyfish even had a protein?

Another interesting coincidence happened when I bought a huge old medical book for ten cents at a thrift store. When I flipped it open for the first time, it opened to a page that mentioned aequorin—the only reference to it in the entire book (which I still have).

Other people might dismiss these "events" as intellectual curiosities or just sheer coincidence, but for me this series of unusual situations affirmed my belief that Quincy Bioscience and Prevagen were meant to be, and that the things we are doing to fight the effects of aging will make a big difference in the quality of our lives as we age.

Keeping Up with the Dream

Having my graduate position evaporate as it did, I took a job with a contract chemical packaging company in Milwaukee, a large operation that packages and develops cleaning products for customers around the world. This gave me a tremendous opportunity for learning about various facets of business, including product development and general management. I remained interested in the jellyfish protein on the side and kept abreast of new developments that were being published in the field of calcium regulation in the aging and disease processes. My idea continued to make sense—it almost felt like an itch I needed to scratch.

In the spring of 2004 I had lunch with Mike Beaman, a friend and business associate who owns Quincy Resource Group, at Applebee's. I shared with Mike my passion for the jellyfish technology. We talked about launching a biotechnology company together. Mike wanted to diversify his investments anyway and was highly intrigued by the incredible potential of aequorin, as well as the enthusiasm I held for this opportunity. That day we sketched out a business plan on a lunch napkin for a new company called Quincy Bioscience. Our goal? To take this jellyfish protein and create a drug to help fight Alzheimer's disease and other related neurodegenerative conditions. We started immediately to plan our venture.

This was actually a very interesting time to start our venture—on June 5, 2004, President Ronald Reagan died of Alzheimer's disease. The president's death brought a tremendous amount of national attention to memory loss, aging, and the ravages of Alzheimer's disease. The official start-up date for Quincy Bioscience was June 7, 2004, just days after the passing of President Reagan.

We knew our aequorin technology could play a very important role in maintaining brain function in older adults, as well as fighting the serious neurological diseases that are associated with a lack of calcium-binding proteins in the brain, such as Alzheimer's. President Reagan had a plaque on his desk that read, "It can be done." Everyone at Quincy took that message to heart, which remains a motivation factor for the entire company.

Our first office space was at Quincy Resource Group's manufacturing complex in Richfield, Wisconsin. We quickly moved from that location to a 1300-square-foot office space within University Research Park in Madison. We also joined forces with Dr. James Moyer, who happened to have just left Yale University to pursue research at University of Wisconsin-Milwaukee. Here is yet another one of those incredible signs—who would have expected that one of the world's top calcium physiologists would relocate to Milwaukee? Dr. Moyer's deep expertise in the field of calcium-mediated dementia and cellular death is extremely rare in the scientific community. Not only was Dr. Moyer now "in the neighborhood," but he had landed a position at my alma mater, where my ideas about aequorin first took root. Our first meeting went well, and shortly afterward we began our research into animals with the jellyfish protein.

Building a Company

When we first started, Quincy Bioscience only focused on the pharmaceutical opportunities that could be developed from aequorin—we knew going through the very complex approval process with the FDA would be a long haul, at least ten years from where we started. This process is ongoing and we are still dedicated to creating pharmaceuticals from our core technology. However, when we received the first laboratory results back, and saw that we had saved a third of the brain cells in our animal tests, we were tremendously excited. Results of this magnitude are extremely rare in the world of research and development. We had kept 30 percent more of the brain cells alive with aequorin, a natural compound that has been proven safe over 30 years of use as a biochemical marker!

We decided to widen our focus beyond producing only pharmaceuticals that would be marketed to physicians. We wanted to help more people, quickly, by getting aequorin into their hands as consumers. After all, if you have Alzheimer's, a breakthrough ten years from now, or even five years, is probably too late. So, we created Prevagen, which is classified as a dietary supplement. We had a safe and natural product that was demonstrating some remarkable results! Time was of the essence in getting this technology to market.

Prevagen became available to consumers on September 1, 2007, and the results of people taking the product have been tremendous. They are especially pleased our technology has been made available as a dietary supplement, which has sped up the delivery of relief rather than having to wait years for the pharmaceutical compound to be approved for specific diseases. People have been able to take a more active part in their own health care by choosing an alternative product that is from a safe and natural source. With many of the current pharmaceutical choices offering limited help in treating neurological illnesses, people want more effective choices that work and we're blessed to have Prevagen help meet that need.

Since we founded Quincy Bioscience in 2004, the science behind our research and development into the mechanisms of age-related and disease-related cellular death is being noticed around the world. We are also working with the best and brightest minds in the biotech industry and academia—combine that with the promise Prevagen holds for improving the lives of so many people, in so many different ways, and it's not hard to smile!

Strong Science

Having repeatable, scientific data is the key to helping people who are interested in natural, alternative health products. Some people say Prevagen sounds too good to be true—until they start looking at the research. Even though many alternative medicine products come from natural sources, they may not have the science to back them up.

We have spent a decade developing this product. Quincy Bioscience is making a major contribution to health by giving consumers a new way to protect their brain cells from memory loss and aging. And it's not just our research—aequorin has been used for a very long time with absolutely no signs of toxicity at any level.

The most important part of my approach to life and business is having faith—faith in something greater than man, faith that something greater guides our paths. That is who I am, that is our mission, and that is how Quincy Bioscience will help the world—one person at a time.

Recapturing Lost Memories

Nancy M. of Menomonee Falls, Wisconsin expressed interest in taking Prevagen™ because dementia runs in her family. "I didn't know what to expect, but what happened to me was miraculous," she said. "Without a doubt it's the most wonderful product I've ever taken."

Nancy, who is 60 and a retired legal secretary, has suffered from migraine headaches since she was first diagnosed by a medical doctor at the age of seven. She and her husband Michael were part of a group that tried Prevagen. "Growing up, my migraine headaches were so bad, sometimes I had to miss half a week a school," she said.

Prevagen has helped Nancy live without the debilitating effects of migraine headaches and without the need for prescription drugs, over-the-counter medicine, or massage therapy, all of which she used for most of her life to combat the grueling affects of her headaches. "One of the most wonderful things Prevagen did for me was something simple, but it was a big benefit," she said. "Prevagen gave me extra time. Migraine sufferers know how difficult it is to function day to day when you're feeling miserable. Because of the pain, you lose time."

Nancy also found she could enjoy being outdoors more than she ever expected. "I can look at the bright lights outside," she said. "That's such a simple thing, but it was difficult for me to do before I took Prevagen."

Perhaps one of the most important things Prevagen did for Nancy was this: "I noticed right away that it gave me my dignity back. Sometimes I have trouble remembering what words come at the end of a sentence. I felt humiliated, but that has all changed now, thanks to this miraculous product."

After Nancy stopped taking her advance samples of Prevagen, her migraine headaches returned. "I can't wait until this becomes available to consumers everywhere," she said. "I'm so impressed with it that I'd stand on a street corner and sell it if I were asked to."

Chapter 2.
Getting Old Doesn't Have to Be a Bummer

We all fear the process of getting old—not being able to do what we could when we were young or even five years ago. The signs and symptoms of aging are numerous and are challenging, if not debilitating, to deal with: not being as strong, agile, quick or even having less endurance throughout the day; becoming confused or uncertain, repeating stories or forgetting where we put things; being the slow driver on the interstate who is holding up traffic; having more health issues, perhaps serious ones; worrying about how to pay for those medical bills, or long-term care insurance. And, perhaps, worst of all, there is the chance of normal aging challenges developing into a devastating illness like Alzheimer's disease—chances are we know someone who is afflicted with it, or a family that has suffered deeply because of it.

Losing independence is one of the greatest fears people have in their "senior" years. The final stages of Alzheimer's slowly strips victims of their memories, dignity, and eventually their lives. It's not an easy way to go.

In fact, physicians and researchers are beginning to think that the early lapses in short-term memory, or fleeting moments of confusion, we experience in our 40s and 50s (which we quickly attribute to stress, multi-tasking, or not getting enough sleep) could actually be the earliest stages of Alzheimer's disease.

The Alzheimer's Association indicates about 26.6 million people worldwide were living with the disease in 2006. Furthermore, it predicts Alzheimer's disease will quadruple by 2050 to more than 100 million, meaning 1 in 85 persons worldwide will be fighting this crippling illness. More than 40 percent of those individuals will be in late-stage Alzheimer's, which requires a very high level of personalized care.

"The number of people affected by Alzheimer's disease is growing at an alarming rate, and the increasing financial and personal costs will have a devastating effect on the world's economies, health-care systems, and families," commented William Thies, vice president of Medical and Scientific Relations for the Alzheimer's Association during the 2007 Alzheimer's Association International Conference on Prevention of Dementia in Washington, D.C. "We must make the fight against Alzheimer's a national priority before it's too late. The absence of effective disease-modifying drugs, coupled with an aging population, makes Alzheimer's the health-care crisis of the 21st century."

That is just Alzheimer's—there are other scary neurological disorders that are striking aging Americans in increasing numbers, including Parkinson's disease, multiple sclerosis, amyotrophic lateral sclerosis (ALS), Pick's disease, and stroke. All these diseases are progressive in nature and impossible to treat effectively; they create huge amounts of emotional and physical trauma and wreck the victim's quality of life. It's estimated that 15 million people in the U.S. suffer from serious neurodegenerative disorders, and thousands more are diagnosed every year, as baby boomers grow older.

Soaring Costs

The cost of health care also continues to spiral out of control. Of course, older Americans have higher health-care costs as these diseases and health conditions take hold.

The average amount spent by Medicare on people with dementia (largely Alzheimer's) is almost three times more than the amount Medicare spends on other types of older patients ($13,207 versus $4,454 per year). Much of this money is devoted to hospital care and nursing-home care. (One reason these costs are so high is that skilled nursing-home care is at least ten times more expensive for patients with dementia.)

Long-term care was not a thought 20 or 30 years ago. Due to Americans living longer and receiving better medical care, combined with the huge increase in lengthy debilitating conditions like Alzheimer's, long-term care policies have become increasingly expensive which, unfortunately, many aging Americans simply cannot afford.

In the United States alone, nearly $2 billion is spent every year on therapeutics for Alzheimer's disease. Combine America with France, Germany, Italy, Spain, the United Kingdom, and Japan, and that number jumps to almost $4 billion per year. Despite these incredible sales figures and the collective hope for positive outcomes, existing medicines are woefully ineffective against Alzheimer's and other neurodegenerative diseases.

The standard approach in medicine is treating the symptoms, not the underlying causes. However—thankfully—we can do both with Prevagen.

Medical research has shown that many of these terrible conditions can be linked to calcium imbalances in the body, which may result from reduced numbers of calcium-binding proteins. This depletion is thought to be a result of aging or the disease process. The job of these incredible, hard-working compounds is to "lock up" excess calcium in the bloodstream and tissue, thereby protecting the nervous system from the negative effects of too much calcium. Our dietary supplement Prevagen uses a special calcium-binding protein called aequorin that does just that—attaches to excess calcium and prevents it from "misbehaving" within certain sensitive cells. This maximizes cellular health, thereby protecting the brain, cognitive function, memory, and neuronal connectivity. Using Prevagen is a great way to regulate calcium in our bodies, whether we already suffer from calcium-related illnesses or simply want to take a proactive approach to maintain the best possible health.

The Role of Calcium in Aging

Most Americans know how important calcium is for healthy bones and teeth. But the role calcium plays goes far beyond the skeleton—calcium (in its ionic form) is one of the most critical components of the human body. It is a key element in the millions of cellular reactions that occur in our bodies every day. Calcium is essential for brain cell health and the efficient transmission of messages throughout the nervous system. It's also important for keeping genes and cells healthy, active, and supple. What many of us do not know is that having the correct amount of calcium in our cells (known to scientists as calcium homeostasis) is important for our bodies to function at peak efficiency. If there is too much calcium in our neurons, critical chemical reactions start to bog down. Over time, this may lead to earlier onset of the physical changes associated with aging, as well as serious diseases like Alzheimer's.

Normally the concentration of calcium in cells is regulated by the actions of calcium-binding proteins (CaBPs), which latch on to excess calcium ions. But as we age, our bodies produce fewer and fewer of these incredibly important proteins, resulting in more ionic calcium floating through our bloodstream and tissues. This extra calcium is particularly troublesome for aging neurons. A number of studies suggest that when these neurons are activated, the calcium interferes with the cell reactions and causes "misfirings" throughout the nervous system—especially in brain regions that are tied to memory, such as the hippocampus.

There is no question calcium-binding proteins are important for regulating the intracellular calcium concentrations in neurons. Studies have also shown neurons that lack certain calcium-binding proteins are less able to handle chemical stressors to their systems—such as too much calcium.

> A decrease in the number of calcium-binding proteins in our bodies can result in too much calcium within our cells, which accelerates the aging process, chronic or acute inflammation, and the onset of serious neurological diseases.

With such an important role in the functioning of the brain, it's not surprising that calcium has been intensely studied in the fields of learning, memory, and aging. Over the last fifteen years, scientists are starting to believe that a breakdown in regulating calcium is a major factor in the development of aging-related learning and memory problems seen in many species, including humans. Research has also shown that calcium-dependent processes are important for associative learning in both adult and aged animals. Other scientists indicate that using compounds that block the influx of calcium, such as calcium channel blockers, actually improves learning and memory. Administering medication to control the rates of calcium entry into neurons has also been proven somewhat helpful in boosting the cognitive thinking of older people.

Many other studies suggest a selective decrease in certain calcium-binding proteins occurs in the brains of aged animals, including humans. Loss of these important proteins with advancing age may leave neurons vulnerable to even moderate increases in the amount of intracellular calcium. The relationship between calcium, neuron degeneration, calcium-binding proteins, and aging suggests that replenishing these proteins in the neurons, through supplementation, especially in parts of the brain that are known to degenerate with advancing age such as the hippocampus, will slow the effects of aging.

Calcium's Link to Disease

With most major neurodegenerative diseases, there is an excess of unregulated calcium in the areas that are afflicted, such as the brain. As these diseases progress and the calcium homeostasis goes awry, the concentration of free calcium ions in the nervous system increases. The body steadily loses its ability to control how this calcium is stored and used. These free-roaming excess calcium ions start to wreak havoc on neurons. The unregulated calcium triggers damaging cellular events that impair and eventually kill the neurons. Disturbances in calcium homeostasis are known to be associated with Alzheimer's disease and other neurodegenerative diseases. Increased total calcium levels, as well as reductions in key calcium-binding proteins, have been found in patients with Alzheimer's, Huntington's, and Parkinson's diseases.

So what can we conclude?

In neurodegenerative diseases there is a known depletion in neuroprotective calcium-binding proteins, which correlates directly with the progression and increasing severity of the disease. This strongly suggests that calcium-binding proteins serve an important role in protecting neurons from calcium overload and keeping the cellular circuitry firing properly. The

Gift From The Sea

easiest way to replenish the calcium-binding proteins that we lose as we age, or fight disease, is with Prevagen.

Protecting Brain Cells with Prevagen™

Quincy Bioscience has patented the nutraceutical and pharmaceutical applications of the calcium-binding protein aequorin. This remarkable compound comes from jellyfish, which have very simple nervous systems. Its natural bioluminescent (glow-in-the-dark) properties when bound to calcium make it ideal for observing changes in calcium ion concentrations, as well as observing calcium uptake by neurons. Over the thirty years aequorin has been in widespread use as a biochemical marker in laboratory research, it has never been used in any therapeutic manner. Quincy Bioscience believes aequorin will be beneficial in treating pathological conditions that result from the disruption of calcium homeostasis in the human body.

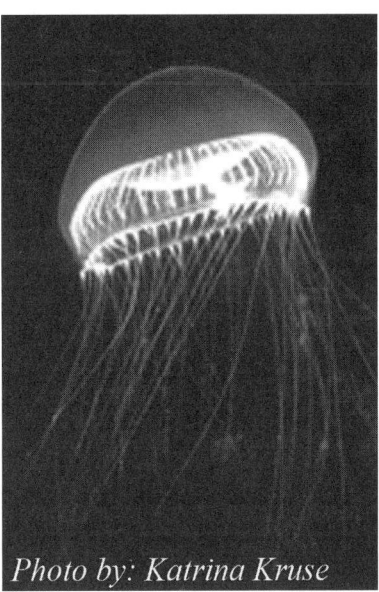

Photo by: Katrina Kruse

Figure 1 Aequorea Victoria Jellyfish

Aequorin comes from a family of calcium-binding proteins that are very similar to those found in humans. Jellyfish have simple nervous systems, and aequorin works within jellyfish to sequester excess calcium ions by binding with them so the calcium can be utilized for predation and self-defense (stinging). Maintenance of this internal calcium balance is critical to the nervous system in jellyfish, as well as in humans.

Aequorin is unique from other calcium-binding proteins and has several distinguishing characteristics: it is nontoxic, it doesn't interfere with internal cellular reactions, and each molecule binds with three calcium ions. Aequorin is also very similar in structure and nucleotide sequence to calcium-binding proteins found in humans, such as calmodulin, parvalbumin, calbindin, and calretinin.

> Aequorin, the active ingredient in Prevagen™, has been used for over 30 years with absolutely no signs of toxicity at any level.

Calcium-binding proteins are found naturally throughout the body and are responsible for keeping cells healthy, and the aging process at bay. They work by buffering ions that could cause cellular damage, and even cell death. Unfortunately, over time we lose the ability to make these valuable, age-fighting proteins at the same rate we did when we were young. This is the problem with aging!

Until Prevagen, there were no effective therapeutics for treating the problem of excess calcium ions and the loss of calcium-binding proteins. Prevagen is the first supplement to fight the aging process by replenishing these powerful proteins, which protect our cells from unhealthy amounts of intracellular calcium. Aging, and advancing diseases, may be counteracted by replacing these depleted proteins.

Sharper Memory, Better Focus

Ava S. has been prone to severe headaches for many years. The busy executive—she's president of a creative services company in New York City—indicated that when she started taking Prevagen she noticed the quality of her life began to change for the better. The changes were significant in numerous ways, including the almost near disappearance of daily headaches.

"For many years, I've had about three to five major headaches a week, sometimes as often as one a day," she said. "But when I started Prevagen, I noticed I wasn't having headaches." During the course of taking the product, Ava experienced just one headache, a sharp decrease in what she had experienced over previous years.

"While I don't understand the science behind it, I think it's too big a coincidence that my headaches nearly disappeared," she stated.

As a busy, multi-tasking person with diverse responsibilities, Ava is the first to admit being spread thin and having trouble focusing on one thing for a considerable amount of time. But with Prevagen, she was able to prioritize and focus with vastly improved concentration.

When she recently bought a new coffee maker, her sharper focus helped her out. "I've had an aversion to anything mechanical because I haven't been able to focus long enough to put something together," she said. "It may not sound like much to many people, but recently, when I took the coffee maker out of the box, I was able to easily focus on the directions. For me, this was truly significant."

Like many people who try to find an unfamiliar location, it can be daunting, even with a map. "I'm notorious for getting lost," Ava said. "I always need to write down directions and even then, I'm not good at finding my way to a new address or location. But after about two weeks of taking Prevagen, something really surprised me. Someone gave me directions to a place I'd never been, and I didn't even have to write down the directions. I was able to memorize them and I didn't get lost."

Ava is looking forward to having Prevagen on the market. In the short time she has taken it she has been impressed by its many benefits.

Interestingly, during the time Ava took Prevagen, her improved quality of life was noticeable to others. "I knew I had a new, calm, focused energy, but several people, who had never heard of Prevagen or knew I was taking it, told me that I looked very rested, happy and calm," she added.

Chapter 3. The Science Behind Prevagen

With Quincy Bioscience firmly established, and a team of supporters in place, it was time to take the next step. We were now ready to embark on the critical scientific experiments and testing of aequorin as a neuroprotective agent.

Years of studying the literature and following the developments in the science of calcium-binding proteins kept me up to date on the latest research involving aequorin and other proteins. Because of my own knowledge, I was already familiar with many of the scientists worldwide who were studying calcium-binding proteins. I began to look for research scientists who would be excellent candidates for collaboration with Quincy Bioscience.

Dr. James Moyer immediately came to mind. His research on the role of calcium in brain cells, specifically aging brain cells, began while he was a graduate student at Northwestern University in Chicago. He continued this research after earning his doctorate and joining a laboratory at Yale University. Dr. Moyer's expertise is on how calcium influences neuronal cells and the brain's ability to learn and remember. Even though calcium is necessary for communication and the processes of learning and memory, too much calcium leads to cell death

I contacted Dr. Moyer in July 2004, shortly after he moved to UW-Milwaukee in Wisconsin. With absolutely no frame of reference for my call, my request for a meeting may have struck him as a little odd. Yet he graciously offered to hear me out and discuss my ideas about testing aequorin in aging brain cells. When I met him and visited his laboratory on the University of Wisconsin-Milwaukee campus, I was further astounded. His laboratory was the same space once shared by the scientist I worked with as an undergraduate at that same institution. These things in life do not happen by accident!

I then introduced Dr. Moyer to my business partner Mike Beaman. As we talked about the experiments that needed to be conducted, and the various strategies we could employ for answering our key questions, we began to forge our relationship. Dr. Moyer later told me that these personal interactions were what convinced him we were serious about developing aequorin as a therapeutic agent. He realized that we were committed to our goals.

In fact, Dr. Moyer told me later he was excited about the concept the first time I mentioned it to him, and that he thought it was worth pursuing—especially since it involved pathways in the brain that he already knew well. He did not have to be sold on the concept—just that the right team was in place. Fortunately for everyone, he liked what he saw and joined the team.

The Testing Begins

We first had to decide which model to use for testing the neuroprotective effects of aequorin. Given Dr. Moyer's years of experience working with rodents, and that rodent models had been yielding reproducible results in these kinds of studies, we decided to start with rats.

We had a wide range of neurodegenerative diseases to choose from for our initial investigation into aequorin's beneficial effects. We decided to focus on stroke, and in particular, ischemic stroke. This type of research is the most readily accepted to assess neuroprotection because the results can be obtained in a short period of time, accurately replicating the type of damage that occurs over decades in the lengthy aging process. Ischemic stroke is characterized by a deprivation of oxygen to the brain, usually caused by a blood clot that blocks a major blood vessel going to the brain. The oxygen loss then triggers a cascade of reactions inside brain cells—many involving calcium and the proteins that bind and control calcium levels—that ultimately kill the affected brain cells.

For years stroke researchers have been aided by a well-developed technique for isolating sections of rat brains and then subjecting them to conditions that mimic ischemic stroke. Because a stroke can occur in virtually any part of the brain, we again faced an important decision as to where in the brain to focus our efforts. This decision was easy—the hippocampus. Not only is this brain structure responsible for forming new memories and the focus of considerable research on learning and memory, but Dr. Moyer has also conducted many studies using hippocampal cells.

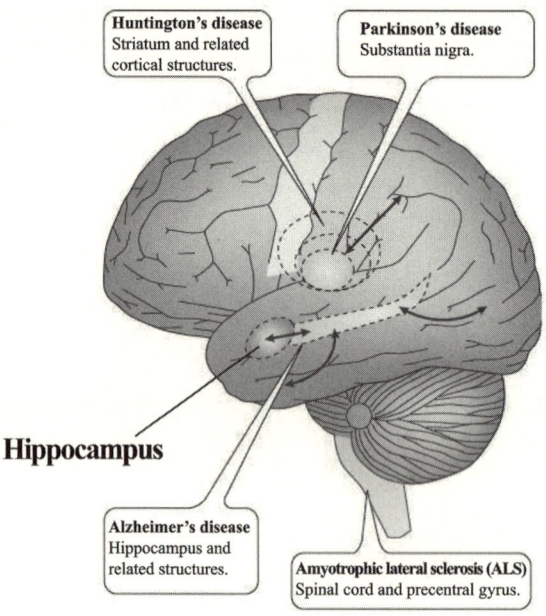

Figure 2 Regions of the brain where aging diseases occur.

We moved forward to test our hypothesis that aequorin would protect brain cells from traumatic injury like stroke. Yet what further interested me was the possible impact aequorin might have on slowing down the aging process. Even in healthy individuals, the mere process of aging causes hippocampal cell activity to falter and slow down. The natural wear and tear on brain cells as we age, and their ability to function at optimal levels, depends in part on having the proper balance of calcium. Was it possible that aequorin could play a role in halting, or at least slowing down, this typically unstoppable process?

Dr. Moyer's research had already demonstrated this critical phenomenon: that animals in middle age experience a decrease in the number of brain cells containing an important calcium-binding protein called calbindin. Dr. Moyer published an article in 2001, where he showed that the amount of calbindin in rat perirhinal cortex cells (another region of the brain involved in memory) starts to decline quite early in the aging process.

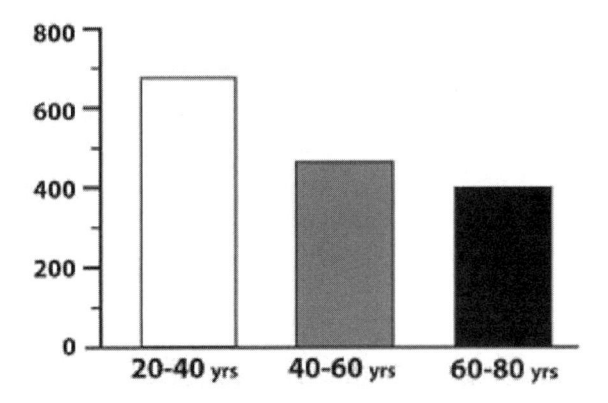

Figure 3 Calcium-Binding Proteins Decrease with Age

Even though these animals were not yet showing signs of memory loss, changes were taking place in their brain cells. What if we could intervene and replenish those diminishing proteins before it's too late and brain function begins to decline?

With our research questions well defined and a reliable experimental model in place, Dr. Moyer and his laboratory personnel were eager to begin. Our first experiments using aequorin were initiated in early 2005.

Dr. Moyer worked with rats and subjected them to conditions that mimicked ischemic stroke. Remember, an ischemic stroke occurs when the blood supply to the brain is suddenly cut off, depriving the brain cells of necessary oxygen. The animals were harvested to obtained brain slices— taking great care to keep the cells alive—and placed them in a chemical solution for a specified amount of time. The solution was depleted of all oxygen (nitrogen gas is substituted in its place), and this well-established scientific procedure allowed us to understand how these cells reacted to such challenging conditions.

Naturally, many cells died when subjected to this stroke scenario. The way we measure how many actually die is by staining them with a chemical dye called trypan blue. In dead cells, the dye is able to infiltrate the cell and stain it blue (seen dark in figure). Healthy, intact, functioning cells, however, remain impermeable to the stain and will not allow it to enter—they remain colorless.

The key to determining aequorin's ability to protect brain cells from death under stroke conditions is comparing the results from rat brain tissues that were injected with aequorin to those not treated with aequorin. Was there a difference in the number of cells that remain alive? The following example from these landmark studies demonstrates aequorin is indeed neuroprotective and protects cells from death when challenged by ischemic stroke:

Control (Untreated Cells)   *Dead Cells*   *Aequorin Treated Cells*   *Healthy Cells*

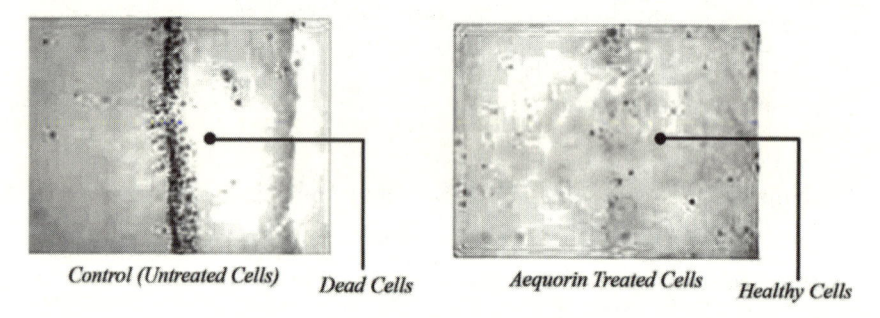

Figure 4 Dead Cells vs. Healthy Cells

When we calculated the numbers we were stunned, and thrilled, to see such a significant effect: Up to 45 percent of the cells treated with aequorin were protected! It's important to note that these experiments were repeated many times, and the results were published and presented at a scientific conference the following year. There was no doubt that aequorin worked!

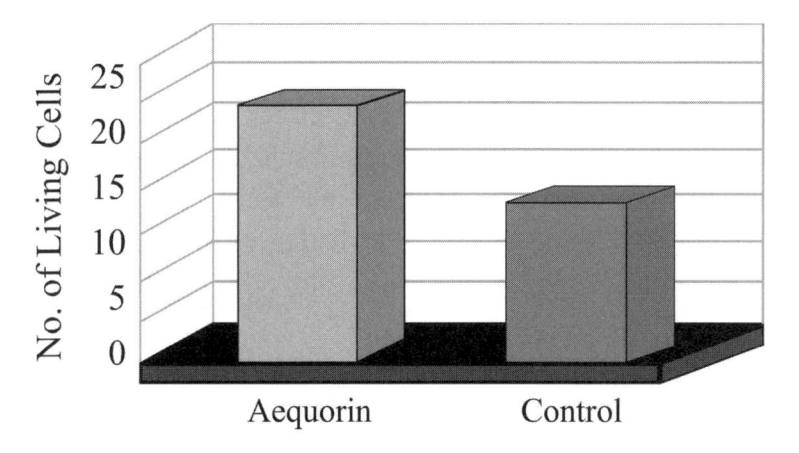

Figure 5 Aequorin Treated Cells

At this point we were convinced aequorin protected cells from death in an ischemic stroke model. But now we needed to answer other critical questions if we wanted to apply what we found to other diseases of aging, such as Alzheimer's and Parkinson's. Was there a way to assess and measure aequorin's neuroprotective effects with respect to the phenomena of memory loss and declining cognitive function as we age?

This critical question formed the basis of more scientific studies, and results continue to be promising. Another advantage of teaming up with Dr. Moyer is his expertise in the field of learning and memory. Using a research technique called Pavlovian trace conditioning—based on the famous experiments by Russian scientist Ivan Pavlov that showed conditioned responses can be taught—we studied how aequorin influences the capabilities of rodents to remember and to learn.

Pavlovian trace conditioning is used to evaluate learning and memory deficits related to aging. As animals get older, is their ability to learn and remember affected? These types of studies are routinely employed for testing new therapies for neurodegenerative diseases such as Alzheimer's. Testing in Quincy Bioscience laboratories is in progress to see how aequorin improves or prevents the memory loss that naturally occurs with aging.

Mark Y. Underwood

History of Aequorin

This story of a gift from the sea would not be complete without some insight into the rich and interesting history behind the discovery of aequorin. For decades this unique protein isolated from jellyfish has been used for research purposes, although Quincy is the first company to change the emphasis and test aequorin for its promising ability to fight aging and disease.

Through its use in scientific laboratories around the world, aequorin has proven to be completely safe and has never shown any dangerous effects. If it were not for the technological advances in large-scale production, such as recombinant protein technology, providing aequorin for health purposes would not have even been possible five to ten years ago. We are all fortunate that such progress has been made, and the quality controls inherent in the process of manufacturing aequorin mean we can now offer it as Prevagen.

> Using state-of-the-art recombinant protein technology, we now have the ability to manufacture large amounts of aequorin, without harming any jellyfish.

But let's go back a half century or so. . . .

Without the tenacity and commitment of a group of scientists at Princeton University back in the 1960s, we might not be reaping the benefits of Prevagen today. It has been almost fifty years since the aequorin protein was first isolated, purified, and given the name aequorin from the genus of jellyfish in which it was first discovered, *Aequorea victoria*. The allure of the protein back then was its ability to illuminate, or light up jellyfish as they floated in the sea. Imagine the thrill of watching these creatures twinkling in the dark waters like stars in the night sky.

Dr. Osamu Shimomura, a Princeton University scientist working under the leadership of senior scientist Dr. Frank Johnson, is responsible for discovering and purifying aequorin in 1962. His achievement involved first traveling across the United States from New Jersey to the Puget Sound in Washington. Along with his wife, a fellow scientist, and Dr. Johnson at the wheel of an old station wagon, the foursome made the trip in seven days. Their destination was Friday Harbor, home of the University of Washington's marine biological laboratory, and a phenomenal source of *Aequorea* jellyfish. The scientists' goal was to determine what caused the jellyfish's brilliant luminescence.

Their scientific mission may have been more than they bargained for. For many painstakingly laborious summers they traveled back to Friday Harbor, collected jellyfish by hand, and used crude techniques to separate the portion that contained the mystery luminescent protein. The team then transported the relevant jellyfish parts back to the east coast and spent the months between summers trying to purify and characterize the protein of interest.

Aequorin was finally described in a research report published in 1962. In the report Dr. Shimomura explained that the protein emits light by binding to calcium inside jellyfish cells. In more scientific terms, each time three individual molecules of calcium bind to one molecule of aequorin, a brief flash of fluorescent light is emitted, causing *Aequorea* jellyfish to luminesce.

While this "trick" is special to the *Aequorea* jellyfish, it turns out that aequorin's ability to bind calcium is a function shared by a large family of similar proteins found in creatures from jellyfish to humans. But aequorin's unique talent of emitting light upon attaching to calcium molecules meant it was a highly valuable tool for scientists who wanted to track the movement of calcium in living organisms. By introducing aequorin into cells of interest, and using scientific tools that allowed them to see the aequorin light up, researchers now had a way of measuring calcium levels. Since then, scientists all over the world have been able to design experiments that rely on this unique property of aequorin to study calcium and its many functions in the body.

Because aequorin binds to calcium inside cells, it is classified as a calcium-binding protein. As mentioned, there are many other calcium-binding proteins (all share a similar structure) that exist in all the cells in our body and thus have the ability to regulate and control biological processes that involve calcium. Calcium is the most abundant mineral in our body and influences many aspects of our general health including brain cell function, bone formation, blood clotting, and wound healing.

Years after the aequorin protein was isolated and purified by Dr. Shimomura, its protein structure or three-dimensional shape was determined. By "seeing" how the protein was folded and contorted in its active state, researchers could better understand how and why it binds to calcium.

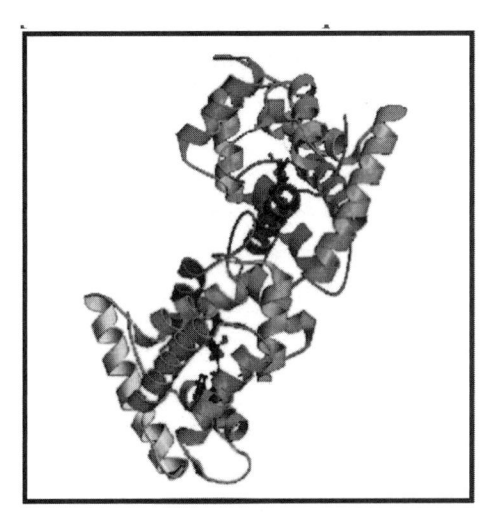

3D Model of Aequorin

Figure 6 3D Model of Aequorin

Dr. Shimomura published his version of the events that led to the discovery of aequorin, beginning with the first time he saw the *Aequorea victoria* jellyfish and ending in the 1980s when he and his colleagues finally established a more efficient way of isolating aequorin for research purposes. The following excerpt from his publication, entitled "A Short Story of Aequorin" (*The Biological Bulletin, August 1, 1995)* illustrates the tremendous progress that has been made—due mainly to his efforts—in the technology necessary to produce large amounts of aequorin:

> We resumed the jellyfish operation at Friday Harbor in the summer of 1967, not anticipating that it would continue for the next 20 years. To collect more jellyfish, we expanded our fishing ground beyond the lab dock, adding the Chevron dock (a small commercial pier), the town dock (public pier), and the shipyard (a covered boat storage), and we used a car to move around and to transport the buckets of jellyfish. When the current carried the stream of jellyfish far beyond the docks, we

also used rowboats to collect jellyfish, a tricky activity that occasionally caused a collector to fall into very cold seawater. The Chevron dock was our favorite place during the first two or three years, because there was a part of it where a large number of jellyfish would stack up on an early morning tide. We had to be careful, however, not to make noise that might awaken sleeping people on the boats.

The town dock was very small—almost nonexistent—in the late 1960s; but then it was rapidly expanded. We harvested jellyfish every morning and evening. The collectors were usually my wife, our son and daughter, a couple of assistants, and me. Dr. and Mrs. Johnson also helped for the first several years. Because jellyfish are nearly transparent in seawater, they cannot easily be seen with untrained eyes. Our children were only three or four years old when they began collecting jellyfish with specially-made short nets; they had become as efficient as an average adult by the age of eight; and through high school they continued to be great helpers in my project.

Before beginning a collection, we filled buckets about half full with seawater and placed them strategically along the edge of pier, then gathered jellyfish until the buckets were completely full. When a dense stream of animals was passing the dock, we could collect at a rate of 5-10 jellyfish per minute. When all the buckets were filled, we poured off some water to about 80% capacity, and then covered each bucket with a plastic bag to prevent seawater from spilling during transportation. The buckets—each crammed with about 100 jellyfish in very little water—were then packed into the trunk of a car (which could accommodate twelve buckets) and rushed to the lab. More buckets were usually transported to the lab on a Boston whaler by one of the assistants. Once at the lab, and before any rings were cut, the jellyfish were kept in aquaria to revive. In this manner, we were able to collect an average of 3,000 to 4,000 jellyfish each day at the town dock.

It once took two tons (4,000 pounds!) of jellyfish to produce a mere 125 milligrams (about one bottle of Prevagen) of aequorin over the course of several months. Today, using recombinant protein technology, we can produce kilogram quantities of aequorin in a matter of days without killing a single jellyfish. What a difference! Gone are the days of catching these slippery, transparent creatures in nets or buckets, lugging them into the trunk of a car, and carrying them back to the lab for further manipulation.

The Safety of Prevagen™

The kind of technology used to produce Prevagen is now commonplace in the manufacturing of many nutraceuticals and over-the-counter dietary supplements. Several different facilities that specialize in recombinant protein production have contracted with Quincy Bioscience, and all meet the highest standards of quality control and safety.

We embarked upon several rounds of testing at the very start of Prevagen production, and the results were just as we expected. Single doses of Prevagen given orally to laboratory rats produced absolutely no toxic or adverse symptoms. All test animals were thoroughly examined for any changes in the tissue pathology or cells of various internal organs, and absolutely no adverse reactions were found.

Mark Y. Underwood

The results of the safety test of Prevagen met our expectations, and we have complete confidence that our product is safe and effective. Test data showed that it is safe to a level of 5,000mg/kg of body weight–this is equivalent to 40,000 capsules ingested at one time by an average 80-kg male. Aequorin has been used in laboratory studies for more than three decades with no adverse reactions, but only until recently could enough be manufactured to demonstrate these higher levels of safety. Scientists using aequorin to measure calcium levels inside cells have long known that it produces no ill effects in experimental animals, and no damage to any cells exposed to aequorin has ever been reported.

Our confidence in Prevagen's safety forms the foundation of our commitment to make this extraordinary supplement available to consumers. It is important to remember that aequorin is a natural calcium-binding protein and that Prevagen will replenish the very calcium-binding proteins that your own body produces, and then loses, as you age.

Clarity of Mind

At a September 2006 conference, Pam M. overheard Mark Underwood talking about the launch of Prevagen and how it protects the body by adding valuable age-fighting proteins. Pam, 44, who has been a consultant for clients in the health and wellness industry for over 20 years, is the marketing director for a consulting firm in New York. "When I heard Mark talk about Prevagen, and how the product helps slow cell death, I was intrigued," she said. "My aunt died from Alzheimer's disease, so I was drawn to learning more about this."

Pam, who was born in 1963 and is at the end of the baby boomer curve, represents many people in their mid-years who know that living longer doesn't always equate to living longer and feeling good. "I wanted to do both—live longer and feel better, and so I saw Prevagen as a natural daily supplement," she said. "It's a natural product, helps me stay healthy, and gives me an excellent opportunity for investing in my future."

Since she had expressed a strong interest in Prevagen, Pam was among those who tried a limited supply for 30 days the summer before its launch. Pam reported that "I didn't expect immediate anti-aging results from taking a month's supply, but I was delighted to take it knowing that I was doing something good for myself."

Pam said she also experienced noticeable results right away. "The number-one thing is that I could handle more things, about twice as many things, and it all stayed clear in my head," she said.

If you're wondering if Pam would recommend it to others she said, "I've already 'sold' it to about 50 people who are anxiously awaiting it. I can hardly wait until Prevagen becomes available this fall. Not only will I buy the product for my 89-year-old mother, it will be a supplement that I will take for the rest of my life."

Chapter 4. The Horizon

It's been a long and illustrious journey on the path to produce Prevagen. What began with an obscure protein found in a little-known jellyfish native to the coastal waters of the Pacific Northwest has evolved into a promising anti-aging dietary supplement, available to millions of consumers. As the international demand for Prevagen intensifies, Quincy Bioscience will ramp up its capacity to reach customers in the United States and the rest of North America, as well as Europe, Asia, and Australia.

As we have said before, our company motto is "It can be done." And when it comes to improving the lives of millions around the globe, you could paraphrase our motto along the following lines: "It *will* be done."

Boomers and Beyond

"No one can avoid aging, but aging productively is something else."

—Katharine Graham, publisher of T

he Washington Post (1917-2001)

Clearly, the release of Prevagen could not come at a better time.

Why?

We are reaching a remarkable stage in history. In the coming years, an unprecedented percentage of the population will be 65 and older.

Indeed, U.S. Census Bureau projections bear this out:

- After the first baby boomers turn 65 in 2011, the older population will mushroom

- The older population in 2030 is expected to be 72 million, more than twice as large as in 2000

- In 2030, older Americans will comprise 20 percent of the total population

- The median age of Americans rose from 22.9 in 1900 to 35.3 in 2000, and is expected to increase to 39 by 2030

But the graying of the population is not unique to the United States—other parts of the globe will eventually be swept by the same aging trends as well. In 2000, 420 million people in the world were 65 and older, representing seven percent of the world's population. By 2030, this number will jump to 974 million. In certain regions of the world, such as Eastern Europe, the growth of the elderly population will be very dramatic, and will have far-reaching implications when it comes to policy-making and social spending. According to the World Bank, between one-fifth and one-quarter of the population will be over 65 by 2025 in a number of Eastern European countries.

For this growing and significant segment of the population, products offering solutions to age-related health and wellness problems are of great value and interest, and could represent a significant improvement in the older generation's quality of life. With its ability to replace calcium-binding proteins that gradually disappear with age, Prevagen offers hope to millions who already are in the 65-and-older demographic, and to those who eventually will join this age group.

In the face of the coming tremendous population boom in the 65-and-older crowd, social spending on the elderly will have to increase in order to meet their needs. In the U.S. the elderly currently account for roughly 13 percent of the population, yet they receive 60 percent of all social spending. In fact, American taxpayers spend three times as much on the elderly than they do on children. Meanwhile in Europe, spending on old age accounts for 42 percent of total social program spending, and it is expected to increase in the future.

It is important to point out that, for a significant portion of those 65 years and older, health spending can be a serious economic burden. Costly prescriptions, medical treatments, and doctor visits tend to increase with frequency as people grow older. Perhaps it comes as no surprise that 77 million older Americans' health-care expenditures include non-traditional solutions, many of which are priced less than "traditional" means.

For countries with aging populations, and for seniors themselves with fixed incomes, Prevagen offers an affordable, effective way to counter the aging process by replenishing calcium-binding proteins. While it obviously is no substitute for prescriptions or visits to the doctor, Prevagen nonetheless can help supplement seniors' investment in their health through its proven neuroprotective properties.

And, in this way, Quincy Bioscience is reaching a wider segment of the aging market much faster than it would if we concentrated first on developing aequorin as a pharmaceutical product. But, as you're about to discover in the following sections, we plan to continue research into potential therapeutic applications of aequorin for the treatment of certain diseases associated with aging.

The Diseases of Aging

"Old age is not a disease—it is strength and survivorship, triumph over all kinds of vicissitudes and disappointments, trials and illnesses."

—Maggie Kuhn, founder of the Gray Panthers (1905-1995)

If you're like most people, when you look ahead to your post-retirement years, different images come to mind. More time to read, travel, garden, or work on your golf game are just a few examples of the leisurely activities you might enjoy in the twilight of your life. But on the flip side of the coin, there are other possibilities that increase in likelihood as you grow older. Namely, different kinds of diseases that tend to prey on people who are older. The following is a brief look at some of these illnesses, their potential causes, and treatments.

Alzheimer's Disease

Alzheimer's disease is a progressive, degenerative illness that gradually destroys a person's memory and ability to learn, communicate, and function on a daily basis. The chances of getting Alzheimer's disease increase with age. In fact, the number of those with Alzheimer's doubles every five years after age 65. Of the estimated five million Americans affected by Alzheimer's, roughly 4.9 million are 65 and older. By the year 2040, nearly 14 million people may suffer from Alzheimer's disease. On average, those with Alzheimer's die about four to six years after having been diagnosed. However, the disease can last from three to as much as 20 years.

Scientists do not know for sure what causes Alzheimer's, although age and family history are known risk factors. The areas of the brain that are most affected by Alzheimer's include the hippocampus, an area deep inside the brain believed to play a key role in memory, and the cerebral cortex, the brain's outer layer which controls language, thinking, and the senses.

No treatment can stop Alzheimer's, although various drugs are used to treat its symptoms.

Parkinson's Disease

Parkinson's disease causes tremors or muscle spasms that can affect the body's muscles and movement. Speech problems, depression, and memory impairment may also result from Parkinson's, which affects nearly 1.5 million Americans. Approximately 60,000 new cases are diagnosed each year in the United States. Parkinson's usually develops after age 65, although 15 percent of those affected are under the age of 50.

What causes Parkinson's disease? Parkinson's results when nerve cells, or neurons, become impaired or die in a part of the brain called the substantia nigra. These neurons produce a chemical called dopamine, which helps coordinate the body's movements. After roughly 80 percent of these neurons are damaged, the signs of Parkinson's disease become evident.

Because Parkinson's disease symptoms are caused by a shortage of dopamine, treatment of the disease typically involves medications that try to replace or mimic dopamine. There is no cure for Parkinson's disease.

> - By the year 2040, nearly 14 million Americans will suffer from Alzheimer's disease.
> - Parkinson's disease affects nearly 1.5 million Americans.
> - The life expectancy of amyotrophic lateral sclerosis (ALS) patients averages between two and five years from the point at which they're diagnosed.
> - About 30,000 Americans have Huntington's disease, and about 150,000 have a 50 percent risk of inheriting it.
> - Forty percent of those afflicted with Pick's disease may have inherited it from a family member.
> - Calcium imbalances in our brain cells, and the loss of calcium-binding proteins, contribute to the onset of many neurodegenerative diseases.

Amyotrophic Lateral Sclerosis (ALS)

Most people know amyotrophic lateral sclerosis (ALS) as "Lou Gehrig's disease," named for the legendary New York Yankee whose brilliant career came to an abrupt, tragic end after being diagnosed. ALS is a progressive disease that spreads rapidly, attacking nerve cells that control voluntary muscles.

Early symptoms include twitching, cramping, and muscle stiffness, and eventually may involve not being able to move one's arms or legs, chew food, or breathe without the help of medical equipment. Up to 30,000 people in the nation have ALS, and every year 5,600 Americans are diagnosed with the disease. It typically affects those between the ages of 40 and 70. The life expectancy of someone with ALS averages between two and five years from the point at which the patient is diagnosed with the illness.

Scientists do not know what causes ALS, and no cure has been found.

Huntington's Disease

Huntington's disease is a progressive, hereditary brain disorder that causes certain brain cells to deteriorate. Uncontrolled movements, mood swings, and balance problems are early symptoms, and as the disease progresses, patients may lose the ability to walk, talk, or think. It is caused by a single abnormal gene, and if one parent has the gene, there is a 50 percent chance it will be passed on to his or her children. About 30,000 Americans have Huntington's, and about 150,000 have a 50/50 risk of inheriting it.

While it doesn't necessarily afflict the elderly in particular, the signs and symptoms of Huntington's usually surface in middle-aged adults. When it appears in younger people, it usually is more severe and progresses more rapidly. There is no cure for Huntington's disease.

Pick's Disease

Pick's disease, also known as frontotemporal dementia, causes gradual shrinking of brain cells as a result of excess protein buildup. It primarily affects the frontal and temporal lobes of the brain, which control speech and behavior. Consequently, personality and behavior changes, along with speech impairment, are hallmarks of the disease. Pick's disease often is mistaken for other illnesses, such as depression, mental illness, or Alzheimer's disease.

As many as seven million Americans may have Pick's disease, which usually affects those between 40 and 60 years of age. Although scientists do not know what causes Pick's disease, 40 percent of those afflicted with it may have inherited it from a family member.

There is no cure for Pick's disease.

Mark Y. Underwood

Remember the Calcium Connection!

What do the above neurodegenerative diseases of the aging all have in common?

According to an emerging theory, it's "the calcium connection." Calcium imbalances in our brain cells—what scientists like to call imbalances in *cellular calcium homeostasis*—are at the root of these devastating disorders. Specifically, the common links between all these diseases include one or all of the following:

- Decreased calcium-binding proteins

- Increased ionized calcium (for example, calcium that freely flows in your blood and is not attached to proteins)

- Increased calpain, a calcium-dependent enzyme that breaks down and consumes other proteins

Calcium is a baseline player in all of these illnesses. But in contrast to the aging process, which gradually leads to higher concentrations of calcium ions and the depletion of calcium-binding proteins, the neurodegenerative diseases described above involve dramatically higher levels of calcium ions and much lower levels of calcium-binding proteins in brain cells.

Researchers already have been investigating ways to deal with these calcium imbalances in brain cells, which lead to neuronal cell death. One approach has involved compounds that block calcium channels. Many people already use calcium channel blockers to treat high blood pressure and prevent the risk of heart attack by relaxing arteries and increasing blood flow. But new studies have involved using calcium channel blockers to prevent excessive amounts of calcium from entering, stressing, and eventually killing brain cells as a result of neurodegenerative diseases. Preliminary findings show this approach has been moderately successful, improving learning in aged animals and restoring certain brain functions to levels found in younger adults.

Scientists at Quincy Bioscience are considering another route. Instead of simply blocking calcium's entry points into brain cells *after* the damage has been done by these neurodegenerative diseases, we are investigating different ways to use aequorin proactively to prevent massive destruction of brain cells due to calcium imbalances associated with these diseases. In other words, we are exploring how to use aequorin to restore calcium levels to their normal levels *before* the damage occurs—something that could be described as a neuroprotective, pre-emptive shield against ALS, Parkinson's, and similar diseases that take a heavy toll on millions around the world.

Taking on the "Silent Killer"

Quincy Bioscience is also researching the use of aequorin for the treatment of "the silent killer"—stroke.

Stroke is the third-leading cause of death in the United States, and the number-one cause of disability among Americans. Strokes typically impact the elderly, but can affect all age groups.

Strokes occur when blood flow to the brain is cut off, due to a blood clot blocking an artery or because of a ruptured blood vessel in the brain. When there is a blockage of blood to the brain, brain cells die because they do not have enough oxygen or nutrients, or simply because of bleeding in or around the brain.

Ischemic strokes, the most common type, involve the blockage of a blood vessel to the brain. The destruction caused by ischemic strokes is massive and swift, like a tsunami wiping out "villages" of brain cells—at a rate of more than 1.9 million brain cells per minute. Another type, *hemorrhagic* stroke, is caused by bleeding in or around the brain. *Transient ischemic attacks* are like "mini-strokes" which happen when the blood supply to the brain is briefly blocked, and within 24 hours the symptoms disappear.

Symptoms can range from weakness or numbness on one side of the body, to difficulty speaking, walking, or seeing. Paralysis and death can also result. Key risk factors include high blood pressure, smoking, alcohol use, high cholesterol, lack of exercise, diabetes, and diets that are high in fat and/or sodium.

So, how might aequorin fit into the treatment of strokes?

For starters, strokes are marked by significant calcium imbalances in brain cells. Specifically, studies show there is a massive influx of calcium into brain cells during strokes. This spike in calcium levels within brain cells leads to the destruction and ultimately the death of the cells. But aequorin, with its calcium-binding properties, possibly could serve as a "shock absorber," bringing stroke-induced high calcium levels in brain cells back down to "normal." With this in mind, we have been conducting in-depth research into aequorin's therapeutic value in treating strokes. With the proper dosage administered in a timely manner, aequorin could successfully protect brain cells from further damage during stroke.

Fighting Inflammation

Researchers are beginning to believe that chronic, low-grade inflammation in the body can lead to a variety of illnesses, including cardiovascular disease, asthma, and autoimmune diseases such as arthritis and multiple sclerosis. For example, University of Southern California researchers have recently discovered a connection between inflammation and a higher risk of developing Alzheimer's disease.

It is likely that the loss of calcium-binding proteins in the human body, a result of aging or advancing disease, creates an overload of calcium ions that contribute toward inflammation and eventual cell death. These processes would then create an impaired immune system and speed the onset of illness or debility. Future studies by Quincy Bioscience will investigate the possible role of Prevagen and its calcium-binding proteins in reducing acute or chronic inflammation levels within the body.

Pain Relief

When Nancy M.'s husband, Michael, said he wanted to try Prevagen, he did so because feeling better was at the forefront of his mind. He also mentioned that Alzheimer's runs in his family.

Michael, 59, a retired government employee, had taken vitamin E, fish oil tablets, and glucosamine for chronic aches in his knees, neck, and back. "I was somewhat skeptical of taking Prevagen at first because I hadn't experienced results with other products," he reported. "I didn't expect Prevagen to have an effect on my knee pain, but it certainly did. Within a couple of days I felt a pleasant warmth in my knees, and after taking Prevagen for just a few days, the aches and pains in my neck, back, and knees were about 80 percent better than before I took the product."

Another pleasant surprise for Michael was getting a better night's sleep. "I was thrilled that I was able to sleep on my back," he said. "That's something I haven't been able to do for years because of back pain."

Michael also says that, in addition to getting relief from pain, Prevagen made him feel "more alive." "It was a surprising result to me, but it is true," he added. "I really felt more alert, sharper, and more relaxed."

Michael is already an avid fan of Prevagen and is eagerly awaiting its arrival on the market. "Someone I know has back spasms and I've told him that as soon as Prevagen becomes available he needs to try it," he said.

The Promise of Aequorin

The implications of aequorin's therapeutic value are almost as far-reaching as the horizon itself. Take strokes, for example. Every year in the United States about 750,000 people get strokes. Globally, strokes are the second-leading cause of death, killing 4.4 million people each year. If aequorin successfully treats even a fraction of the millions who annually succumb to strokes, it would be a major medical victory. Beyond the potential savings in human costs, there are potential economic cost savings as well. In this country, strokes account for annual losses of up to $62.7 billion in medical treatment expenditures and lost productivity. In view of these statistics, aequorin could potentially save millions of dollars in treatment and lost productivity each year in the United States alone.

Given the tremendous human and economic costs associated with other neurodegenerative diseases, aequorin offers even greater promise and hope. We will continue to invest significant time and effort into the clinical trials required for Federal Drug Administration approval of aequorin as a pharmaceutical treatment for neurodegenerative diseases. That is why our laboratory research is in full swing, in order to bring pharmaceutical aequorin products to market by the year 2014. Thankfully, we already are "ahead of the game," so to speak, given that the basic properties of aequorin have been proven over the past 30 years. As of 2007 we have already reached the eight-year mark in the 15-year cycle (five years ahead of schedule!) it typically takes to develop new drugs, due to the vast amount of existing research on aequorin.

> "When people ask me, as the president and founder of Quincy Bioscience, what experiences I've had taking Prevagen, I tell them it has changed my life in many ways, but here's one example that everyone can relate to," says Mark Underwood.
> "Before taking Prevagen, I needed to write down the grocery list when my wife called and asked me to pick up items at the store on my way home. But I'm not doing that anymore. Prevagen has given me the freedom from writing down the grocery list! That's just another added value that jellyfish have given us as a gift from the sea."

Quincy Bioscience is committed to understanding and treating neurodegenerative diseases, and bringing novel therapeutics to the marketplace. The company expects that the continued development of our protein-based technology will have applications toward the treatment of stroke and neurological-based diseases such as Alzheimer's and ALS. We may also discover that balancing cellular calcium homeostasis by replenishing calcium-binding proteins will help conditions such as cystic fibrosis, viral hepatitis, arthritis, and other chronic inflammatory disorders.

It is our belief that both cellular dysfunction and cellular death can be averted, or at least improved, when intracellular calcium imbalances are managed through a therapeutic approach. By tying up the uncontrolled release of calcium ions with Prevagen, the toxic effects of excess calcium are avoided and cells can continue to function in their normal way. This will have huge implications for reducing the cost of health care for the aging population, since increases in the amounts of intracellular calcium ions can be directly correlated to the aging process of the brain, skin, heart, and other major organs of the body.

At Quincy Bioscience we continue to develop future pharmaceutical applications, which will be marketed through strategic partnerships with larger groups, such as pharmaceutical companies. Two new labs at the University of Wisconsin in Madison are now partnered with Quincy to conduct research regarding aequorin's role in managing stroke disease, as well as the chronic inflammation that is found in arthritis and autoimmune diseases. We have established a highly-qualified board of directors and a scientific advisory panel to guide us as we continue to grow toward the future.

Quincy Bioscience researchers believe that by helping the body manage intracellular calcium, natural aging can be slowed down to increase longevity and enhance the productivity of human life. As we work toward developing these near-future pharmaceutical agents, the public can still benefit from aequorin through our dietary supplement Prevagen. It's all-natural, safe, reasonably priced, and scientifically proven to be effective in keeping brain cells alive. This remarkable compound from the sea, aequorin, is truly a gift to everyone who want to fight aging and live longer, healthier, happier lives.

Selected References

Adams, R.D., and M. Victor. *Principles of Neurology.* New York: McGraw-Hill Book Company, 1985.

Alexianu, M. et al. (1994). The role of calcium-binding proteins in selective motoneuron vulnerability in ALS. *Annals of Neurology* 36(6):846-858.

Appel, S.H. (1993). Excitotoxic neuronal cell death in ALS. *Trends in Neuroscience* 16: 3-5.

Baimbridge, K. et al. (1992). Calcium-binding proteins in the nervous system. *Trends in Neuroscience* 15(8):303-308.

Blinks, J. (1990). Use of photoproteins as intracellular calcium indicators. *Environmental Health Perspectives* 83:75-81.

Detert, J.A. et al. (2006). Neuroprotective effects of aequorin on hippocampal CA1 neurons following ischemia. Scientific abstract, *Neuroscience*.

Disterhoft, J.F., R.J. Moyer, and L.T. Thompson (1994). The calcium rationale in aging and Alzheimer's disease. Evidence from an animal model of normal aging. *Annals of the New York Academy of Sciences* 747:382-406.

Duncan, C. (1987). Role of intracellular calcium in promoting muscle damage: a strategy for controlling the dystrophic condition. *Experientia* 34:1531-1535.

Duthie, G.G. and J.R. Arthur. (1993). Free radicals and calcium homeostasis: relevance to malignant hyperthermia? *Free Radical Biology and Medicine* 14(4):435-442.

Eisen, A. (1995). ALS is a multifactorial disease. *Muscle-Nerve* 18(7):741-752.

Fleckenstein-Grun, G. and A. Fleckenstein (1991). Calcium—a neglected key factor in arteriosclerosis. The pathogenic role of arterial calcium overload and its prevention by calcium antagonists. *Annals of Medicine* 23(5):589-599.

Gibson, G. and C. Peterson (1987). Calcium and the aging nervous system. *Neurobiology Aging* 8:329-343.

Halliwell, B. (1989). Oxidants and the central nervous system: some fundamental questions. *Acta Neural Scandinavia* 126:23.

Hartmann et al. (1994). Disturbances of the neuronal calcium homeostasis in the aging nervous system. *Life Science* 55(25-26):2011-2018.

Heizmann, C. and K. Braun (1992). Changes in Ca(+2)-binding proteins in human neurodegenerative disorders. *Trends in Neurosciences* 15(8):259-264.

Iacopino, A., and S. Christakos (1990). Specific reduction of calcium-binding protein (28-kilodalton calbindin-D) gene expression in aging and neurodegenerative diseases. *Proceedings of the National Academy of Sciences USA* 87:4078-4082.

Iacopino, A., and S. Christakos (1984). Specific alterations in calcium-binding protein gene expression in neurodegenerative diseases. Abstract. UCLA Symposium of Neurodegenerative Diseases.

Ibarreta, D. et al. (1997). Altered Ca(2+) homeostasis in lymphoblasts from patients with late-onset Alzheimer's disease. *Alzheimer Disease Association Disorders* 11(4):220-227.

Ichmiya, Y. et al. (1988). Loss of calbindin-D28k immunoreactive neurons from the cortex in Alzheimer-type dementia. *Brain Research* 475:156.

Imbert, N. et al. (1995). Abnormal calcium homeostasis in DMD myotubes contracting in vitro. *Cell Calcium* 18:177-186.

Ince, P. et al. (1993). Parvalbumin and calbindin D28k in the human motor system and in motoneuron disease. *Neuropathology Applied Neurobiology* 19(4):291-299.

Inouye et al. (1985). Cloning and sequence analysis of DNA for the luminescent protein aequorin. *Proceedings of the National Academy of Sciences USA* 82:3154-3158.

Khachaturian, Z.S. (1994). Calcium hypothesis of Alzheimer's disease and brain aging. *Annals of the New York Academy of Sciences* 747:1-11.

Khachaturian, Z.S. (1989). Introduction and overview: calcium, membranes, aging and AD. Z.S. Khachaturian, C.W. Cotman, and J.W. Pettigrew, eds. *Annals of the New York Academy of Sciences* 568:1-4.

Lally, G. et al. (1997). Calcium homeostasis in aging: studies on the calcium binding protein calbindin D28k. *Journal of Neural Transmission* 104(10):1107-1112.

Lamm, Richard D. and Robert H. Blank. The challenge of an aging society. *The Futurist,* July-August, 2005.

Landfield, P.W. et al. (1992). Mechanisms of neuronal death in brain aging and AD: role of endocrine-mediated calcium dyshomeostasis. *Journal of Neurobiology* 23:1247-1260.

Leslie, S.W. et al. (1985). Reduced calcium uptake by brain mitochondria and synaptosomes in response to aging. *Brain Research* 329:177-183.

Lipton, S.A. and S.B. Kater (1989). Neurotransmitter regulation of neuronal outgrowth, plasticity, and survival. *Trends in Neurological Sciences* 12:265-270.

Mattson, M.P. (1992). Calcium as a sculptor and destroyer of neural circuitry. *Experimental Gerontology* 27:29-49.

Mattson, M.P. (1989). Cellular signaling mechanisms common to the development and degeneration of neuroarchitecture. *Mechanisms of Ageing and Development* 50:103-157.

Mattson, M.P. and B. Cheng. (1993). Growth factors protect neurons against excitotoxic/schemic damage by stabilizing calcium homeostasis. *Stroke* 24 (Supplement 1):36-40.

Mattson, M.P. et al. (2000). Calcium signaling in the ER: its role in neuronal plasticity and neurodegenerative disorders. *Trends in Neurological Sciences* 23(5):222-229.

McLachlan, D. et al. (1987). Calmodulin and calbindin D28k in Alzheimer's disease. *Alzheimer Disease and Associated Disorders* 1(3):171-179.

Meyer, F.B. (1989). Calcium, neuronal hyperexcitability and ischemic injury. *Brain Research Brain Research Review* 14(3):227-243.

Morrison, B.M. et al. (1998). Determinants of neuronal vulnerability in neurodegenerative diseases. *Annals of Neurology* 44:S32-S44.

Mouatt-Progent, A. et al. (1994). Does the calcium-binding protein calreticulin protect dopaminergic neurons against degeneration in Parkinson's disease? *Brain Research* 668(1-2):62-70.

Moyer, J.R., Jr., S.M. Kelsey, J.P. McGann, and T.H. Brown (2001). Morphology and distribution of calbindin-D28k in adult and aged rat perirhinal cortex. *Society for Neuroscience Abstracts* 27, program number 327.7.

Moyer, J.R., Jr., J.M. Power, L.T. Thompson, and J.F. Disterhoft (2000). Increased excitability of aged rabbit CA1 neurons after trace eyeblink conditioning. *Journal of Neuroscience* 20:5476-5482.

Mukesh Chawla, Gordon Betcherman, and Arup Banerji. *From Red to Gray: The Third Transition of Aging Populations in Eastern Europe and the Former Soviet Union.* Washington, D.C.: World Bank, 2007.

Nixon, R. et al. (1994). Calcium-activated neutral proteinase (calpain) system in aging and AD. *Annals of the New York Academy of Sciences* 747:77-91.

Pascale, A. and Etcheberrigaray, R. (1999). Calcium alterations in Alzheimer's disease: pathophysiology models and therapeutic opportunities. *Pharmacology Research* 39(2): 81-88.

Rami, A., and J. Kriglstein. (1994). Neuronal protective effects of calcium antagonists in cerebral ischemia. *Life Sciences* 55(25-26):2105-2113.

Ripova, D. et al. (2004). Alterations in calcium homeostasis as a biological marker for mild Alzheimer's disease. *Physiological Research* 53:449-452.

Scharfman, H.E. and P.A. Schwartzkroin (1989). Protection of dentate hilar cells from prolonged stimulation by intracellular calcium chelation. *Science* 246:257-260.

Seto-Ohshima A., P.C. Emson, E. Lawson, C.Q. Mountjoy, and L.H. Carrasco (1988). Loss of matrix calcium-binding protein-containing neurons in Huntington's disease. *Lancet* 1(8597):1252-1255.

Shanahan, C. M. et al. (1994). High expression of genes for calcification-regulating proteins in human atherosclerotic plaques. *Journal of Clinical Investigation* 93(6):2393-2402.

Shaw, P. et al. (2000). Molecular factors underlying selective vulnerability of motoneurons to neurodegeneration in ALS. *Journal of Neurology* 247 (Supplement 1): 17-27.

Shimomura, O. A short story of aequorin. *The Biological Bulletin,* August 1, 1995.

Shimomura, O., F.H. Johnson, and Y. Saiga (1963). Further data on the bioluminescent protein, aequorin. *Journal of Cellular and Comparative Physiology* 62:1-8.

Shimomura, O., F.H. Johnson, and Y. Saiga (1962). Extraction, purification and properties of aequorin, a bioluminescent protein from the luminous hydromedusan, Aequorea. *Journal of Cellular and Comparative Physiology* 59:223-239.

Shimomura, O. and F.H. Johnson (1978). Peroxidized coelenterazine, the active group in the photoprotein aeqourin. *Proceedings of the National Academy of Sciences USA* 75: 2611-2615.

Shinichi, Iwasaki et al. (2000). Developmental changes in calcium channel types mediating central synaptic transmission. *Journal of Neuroscience* 20(1):59-65.

Siesjo, B.K. et al. (1989). Calcium, excitotoxins, and neuronaldeath in brain. *Annals of the New York Academy of Sciences* 568:234-251.

Strazzullo, P. et al. (1986). Altered extracellular calcium homeostasis in essential hypertension: a consequence of abnormal cell calcium handling. *Clinical Sciences* 71(3): 239-246.

Sutherland, M. et al. (1993). Reduction of calbindin-28k mRNA levels in Alzheimer as compared to Huntington hippocampus. *Brain Research Molecular Brain Research* 18(1-2):32-42.

Underwood, M. *The Aequorin Hypothesis—Calcium-Binding Proteins and Disease.* A research monograph. Self-published, 1996.

Verity, M.A. (1992). Ca(2+)-dependent processes as mediators of neurotoxicity. *Neurotoxicology* 13:139-148

Yamada, T. et al. (1990). Relative sparing of Parkinson's disease of substantia nigra dopamine neurons containing calbindin D28k. *Brain Research* 526:303-307.